Jéssica Silva Dos Reis
Jales T. Chaves Filho
Marcos A. Pesqueiro

Morphological description of the tree Triplaris gardneriana Wedd

Jéssica Silva Dos Reis
Jales T. Chaves Filho
Marcos A. Pesqueiro

Morphological description of the tree Triplaris gardneriana Wedd

(Polygonaceae, Caryophyllales)

ScienciaScripts

Imprint
Any brand names and product names mentioned in this book are subject to trademark, brand or patent protection and are trademarks or registered trademarks of their respective holders. The use of brand names, product names, common names, trade names, product descriptions etc. even without a particular marking in this work is in no way to be construed to mean that such names may be regarded as unrestricted in respect of trademark and brand protection legislation and could thus be used by anyone.

Cover image: www.ingimage.com

This book is a translation from the original published under ISBN 978-613-9-64547-3.

Publisher:
Sciencia Scripts
is a trademark of
Dodo Books Indian Ocean Ltd. and OmniScriptum S.R.L publishing group

120 High Road, East Finchley, London, N2 9ED, United Kingdom
Str. Armeneasca 28/1, office 1, Chisinau MD-2012, Republic of Moldova, Europe
Printed at: see last page
ISBN: 978-620-7-74373-5

"The task is not so much to see what no one has seen, but to think what no one has yet thought about what everyone sees."

(Arthur Schopenhauer)

Thank you

I would firstly like to thank God, who has illuminated my path throughout this journey. For the opportunity to be part of the Biological Sciences family, which with great love and affection welcomed me into its walls and arms, helping to form and build the student, woman and future professional that I will be.

This family, like all others, is full of problems, conflicts and challenges, but they are all solved with respect, complicity and friendship, factors that strengthen and build the fraternal bonds between its members. It goes beyond the chairs, tables and walls of a traditional classroom, breaking down boundaries, dogmas and the walls of a journey.

With all my admiration, respect and sincerity, I would like to thank all my friends, colleagues and teachers who have been an integral part of my personal and academic development over the last four years. In particular, I would like to thank my teachers and friends Jales, Everton, Marcos and Daniela, who motivated me, supported me and contributed to the construction and development of this work.

I would like to thank Prof Jales Teixeira Chaves Filho for his guidance, for the advice that undoubtedly enabled me to complete this course, for all the time he took out of his busy life, for his patience and dedication to teaching and for his priceless friendship. For the exemplary teacher, for the motivation in difficult times, for the support in pursuing the fulfilment of my dreams, for being more than the pillar of my education, for being an example of a professional.

To NEA - Novas Edições Acadêmicas, who lovingly and patiently revised the manuscript and contributed to a better understanding of the message we wanted to spread.

I would like to thank everyone who, directly or indirectly, played a part in my education.

SUMMARY

The aim of this study was to describe the morphological characteristics of seedlings and achenes, to identify the storage compounds in Triplaris gardneriana Wedd. diaspores, as well as to obtain information on the physiological aspects of the tree's leaves by means of linear regression equations, the biometric measurements of the length and width of the leaves of the species, determining an estimate of the leaf area, aiding in the botanical identification of this species. The analyses were carried out in the Ecology laboratory at the State University of Goiás - Câmpus Morrinhos, and the morphology of the achenes and young plants in the initial growth phase was assessed. This characterisation was carried out using analytical keys and observations using an electron microscope and magnifying glass. To determine the biometry of the young plants, leaf area was estimated by obtaining the linear equation of the leaves of T. gardeneriana Wedd. To assess the histochemical compounds of the main seed storage compounds, microscopic photographs were recorded and observed. The results indicated that the fruit is of the achene type, the seeds are bitegumented, the seedlings are phanerocotyledonous with epigean germination. The leaves are hypostomatic, with paracytic stomata, mucronate, obtuse base, entire edge. It was also observed that the seeds of the pau-formiga do not have a lipid substance, but have starch as their reserve compound. From the results obtained, it was possible to describe and characterise the species Triplaris gardneriana Wedd. as a plant with bitegumented achenes, ruminate and starchy endosperm. The seedlings are epigeous and phanerocotyledonous with photosynthesising membranous cotyledons. The leaves at the initial stage of development are hypo-stomata typical of mesophytic plants, with paracytic stomata, a mucronate apex, obtuse base and entire edge. The equation for estimating leaf area showed a good fit in relation to the real data and could replace the use of equipment for this purpose. The Triplaris gardneriana Wedd. tree is a neutral photoblastic species that shows high viability and germination potential with rapid speed, demonstrating performance and physiological quality in its biological, morphological and metabolic characteristics.

Keywords: Leaf morphology, achene anatomy, leaf limb biometry.

Summary

CHAPTER 1

Introduction

The Cerrado biome has the second greatest biodiversity in Brazil and is home to many endemic plant and animal species (DURIGAN, et.al. 2011). According to the MMA (2006), the Cerrado has a large territorial extension that is home to a wide variety of heterogeneous ecosystems and 5% of the planet's biodiversity. However, in recent years the states that make up this biome, such as Goiás, Mato Grosso, Mato Grosso do Sul, the Federal District, as well as part of the states of Tocantins, Minas Gerais, Bahia, Maranhão, Piauí, São Paulo, Paraná, Rondônia, Roraima and Amapá, have suffered from direct human action, causing a reduction in diversity and natural resources through burning, deforestation and territorial expansion for agriculture, farming and urbanisation (DINIZ, 2006; BATISTA, 2009).

The cerrado is made up of different phytophysiognomies that make up the grassland, savannah and forest formations, comprising a variety of environments, habitats and physiognomies (SILVEIRA, 2010). Savannas include the cerrado sentido restrito (or stricto sensu), cerrado park, palm grove and vereda (MMA, 2009). Pessoa (2014) points out that the cerrado sentido restrito is divided into cerrado denso, cerrado típico, cerrado ralo and cerrado rupestre, but that in general the physiognomic formations show randomly distributed tree cover, with different densities and shrubs and herbs spaced out in the vegetation.

According to Silva (2007), the cerrado park has trees that cover small elevations in the terrain, known as "murundus". The palm grove is a vegetative type that features a diversity of palm species such as babassu, macaúba and buriti, which is very similar to the landscape of the veredas, which is basically made up of Mauritia vinifera palms and a grassy layer all year round (BASTOS, 2010).

Forests include the cerradão, riparian forest, dry forest and gallery forest formations. The cerradão has sclerophyllous trees, with species from the cerrado sentido restrito and the woodlands, as well as shrubs, herbs and some grasses in the understory (SILVA-JUNIOR, 2012). Riparian forests, in turn, are tree formations found on the banks of water reservoirs (RIBEIRO et.al. 2012). Gallery forests, like riparian forests, are formations located in the beds of rivers and streams, but they consist of a "tunnel" where the treetops of the two banks meet, a phenomenon that is not evident in riparian vegetation types (FELFILI et al., 2000). Santiago et al. (2005) emphasise that these communities are considered by many scholars to be one of the formations with the greatest diversity and richness of species in the biome. Dry forests, on the other hand, are made up of seasonal deciduous forests with different levels of deciduousness (SANTOS-DINIZ, 2011; SANTOS et. al. 2006).

This high diversity and biological abundance of species previously presented by the different types of phytophysiognomies, in addition to the speed with which the biome is being destroyed, has led to the Cerrado being included in the priority areas for worldwide conservation, the "hotspots", thus receiving protection from environmental legislation (OLIVEIRA E MARTINS, 2002). This makes the need for effective measures to preserve and conserve the Cerrado unquestionable (MMA, 2007). Bearing in mind that currently the percentage of species threatened with extinction is greater than the resources available for their conservation (MYERS et. al. 2000).

According to Pinheiro and Kury (2008), environmental conservation only occurs when the components (ecosystems, communities and species) are in good condition. Therefore, in order for the environment to function properly, it is important to first identify and understand the morphological and physiological aspects that influence the behaviour and ecological interactions of a population in order to then understand the functioning of these components and successively protect and conserve a biome.

Thus, when it comes to botanical studies, the description of the morphological structures of seeds and seedlings, as well as the histochemical factors (lipids, starch, carbohydrates, amino acids and proteins) presented by the embryos, have the function of taxonomic and anatomical characterisation of a family, genus or species. Araújo-Neto et. al. (2002) and Silva et. al. (2010) point out that this information also contributes to germplasm bank research, species identification and the natural regeneration of degraded areas, as well as enabling the characterisation of plants still in the vegetative phase.

However, there is little information in the literature about the endemism of genera native to the Cerrado biome, such as those belonging to the Polygonaceae family, despite the fact that the number of species in this family is relatively small in the Cerrado (MMA, 2003). According to Anjen et al. (2003) and Cruz et al. (2008), the Polygonaceae family includes monoecious or dioecious plants, herbaceous annuals or perennials, shrubs or trees.

This group is often used for construction (agricultural tools, rafters, laths, beams, coal, boards, carpentry, joinery and furniture making), such as the species Ruprechtia exploratriscis Sandwith, Coccoloba mollis Casar, Ruprechtia laxiflora Meisn and Coccoloba rosea Meisn and also for landscaping and reforestation due to their beauty during flowering and their rapid growth as can be expressed by some species of the Triplaris genus such as the trees T. americana L., T. weigeltiana (Rchb.) Kuntze and T. Gardneriana Wedd (LORENZI, 2009; 2013; 2014).

In the case of species of the Triplaris genus, as well as these plants having ornamental and reforestation characteristics, they also have a myrmecophilous association. Myrmecophily defines relationships between ants and plants, which can be an interaction for food or shelter (DÁTTILO et al., 2009). The trunks and branches of these trees are hollow, so ants such as Pseudomyrmex triplarinus establish symbiotic and mutualistic interactions, as they live in the trunk of the plant and in return protect it from herbivores (HADDAD-JUNIOR et al., 2009).

Ward (1999) points out that T. gardneriana Wedd., popularly known as pau-formiga, was given this name due to the large number of reports of ant attacks and their interaction. There are few studies in the literature dealing with morphological studies of this species, which could contribute to taxonomic identification and an understanding of the importance of the ant-wood in ecological interactions in the biome, as well as its conservation and preservation. Although some authors have already studied it, there are still morphological and histochemical characteristics unknown to researchers.

Figure 1- Morphological aspect of the tree T. gardneriana Wedd. in the adult stage, located in the urban area of the city of Morrinhos - Goiás; (A) yellow flowers, 5-6 mm in diameter, gathered in a staminate inflorescence 5-18 cm long and supported by stems 10 mm long each; (B) stem 4-7 m high, with a globose crown, with a straight-branched trunk, covered in thin, smooth, peeling bark in thin greyish plates, 20-30 cm in diameter. (C) and (D) morphology of the helix of the fruit dispersed by anemochory.

The species T. gardneriana Wedd. is a dioecious, xerophytic tree with a low, sparse globe-shaped crown, a trunk that is sometimes twisted and branched, covered in smooth, thin, flaking bark (Figure 1) (CASTRO and CALVACANTE, 2010). Its leaves are simple, membranaceous with an elliptical limb, acuminate base, entire and hairy edge, cuspidate apex and alternate phyllotaxis (REIS, 2013). Its paniculate inflorescence has either male or female flowers, producing achene-type fruits, hairy with a helix (Figure 1) (LORENZI, 2013; SILVA and LEMOS, 2002).

The pau-formiga can be found in the cerradão along riparian forests, gallery forests of semi-deciduous broadleaved forests and dry forests (REIS and CHAVES, 2014; ENBRAPA, 2005; MENDONÇA et al., 2005). Its size can vary from one region to another, reaching a height of approximately 4 to 15 metres and a diameter of 20 to 30 cm (SILVA and LEMOS, 2002; CASTRO and CALVACANTE, 2010).

According to Judd et al. (2009) morphological characters are attributes that provide most of the information used to identify plants and build hypotheses about phylogenetic relationships between groups, as well as providing evolutionary evidence.

The aim of this study was therefore to describe the morphological and physiological characteristics of seedlings and achenes, to identify the storage compounds in Triplaris gardneriana Wedd. diaspores, as well as to obtain information on the physiological aspects of the tree's leaves by means of linear regression equations, the biometric measurements of the length and width of the leaves of the species, determining an estimate of the leaf area, aiding in the botanical identification of this species.

CHAPTER 2

MATERIAL AND METHODS

According to Lorenzi (2002), the T. gardneriana species has an achene fruit as a diaspore (dispersal unit). The achene is characterised by being a small, indehiscent, dry fruit with only a thin, adpressed wall (Judd et al. 2009), containing a single seed inside. Therefore, throughout this work the term achene will be used to define the dispersal unit of the species.

The analyses were carried out at the Ecology laboratory of the State University of Goiás - Morrinhos Campus, and were divided into two phases: the morphological analysis of the achenes and plants in the initial growth phase and the histochemical identification of the main seed storage compounds.

In order to determine the morphological characteristics of the internal and external structures of the diaspores of the T. gardneriana species, 50 randomly selected achenes from lots collected in the urban area of the municipality of Goiânia in 2014 at different locations were used. The achenes were then mechanically scarified using water sandpaper and soaked in 30 ml of distilled water in a B.O.D. chamber at room temperature (24°C) for 24 hours under constant lighting. After soaking, the samples were observed using a magnifying glass and microscope to describe the colour, texture and consistency of the integuments, shape, dimensions, position and shape of the hilum, micropyle, raphe and embryo, i.e. cotyledons and hypocotyl-radicle axis.

The development of the embryo during the germination phase, as well as the utilisation of the reserve material in the achenes, was monitored for 31 days, where the results were recorded using microscopic photographs using Leica DM 500 equipment with an attached microcomputer. After germination, the morphological characterisation of the seedlings of the species studied was carried out over 83 days.

For the morphological characterisation of the young plants, 50 germinated plants were selected under laboratory conditions (room temperature and constant light) and then planted in plastic bags (15 cm in diameter and 15 cm high) containing organic substrate. The seedlings were kept in shade (50%) in an area of the State University of Goiás - Morrinhos Campus, receiving regular weekly irrigation (3 times) for 3 years, corresponding to the period from 2012 to 2014 when they were sampled. After this period, the plants were randomly sampled to assess the morphology of the leaf limb, which was characterised using analytical keys.

To determine the biometrics of the plants in the initial growth phase, diameter measurements were taken using a caliper and leaf area was also estimated by obtaining the linear equation of the leaves of T. gardeneriana Wedd. using the methodology described by Benincasa (2003). Expanded leaves from 30 young plants germinated in the Ecology laboratory at UEG-Câmpus Morrinhos in 2012 were used.

To construct the straight line equation, the biometric measurements of the 150 leaves sampled were used, in relation to the result of the length and width of the leaf and its area obtained by weighing the leaf contours. Bearing in mind that the resulting product of length and width provides an area value, which does not correspond to perfect geometric figures, so to obtain the leaf area, a correction factor adjusted according to the area estimate must be used.

leaf using the contour method.

The correction factor was obtained using the statistical regression indices from the straight line equation using the product of length and width as the x variable and leaf area as the y variable. The result of the statistical regression analysis will provide the a and b indices corrected for the difference between the linear measurements and the leaf outline. With the indices determined, it will be possible to calculate the leaf area, using $Y=a+b.x$ provided by the equation of the line. Based on the equation obtained, simply detail the linear dimensions of the leaves and replace the result with the x value from the equation to obtain the leaf area (Y), once the a and b indices have been obtained for the species. Once the estimated leaf area had been obtained using the equation, the results were compared with the real leaf areas, and statistical analysis was carried out using the chi-square method at a 5% probability level to check the significance of the equations obtained.

To assess the reserve histochemical compounds in the diaspores, 50 achenes previously collected in the municipality of Goiânia-GO between August and September 2014 were used and stored in a refrigerator at 5° C in the Ecology laboratory of the Morrinhos-Goiás university unit until they were handled.

As the tissues of the parts analysed had a low moisture content, the plant material for the histochemical observations of starch and lipids was soaked in water for 24 hours and then kept in a B.O.D. thermostatic chamber for the necessary methodological procedures. After soaking, the material was subjected to freehand anatomical cutting using a razor blade.

In order to test for the presence of starch in antwood seeds, we used the lugol reagent histochemical technique described by Macêdo (1997), which consists of placing the plant material to be analysed on a glass slide, then add a drop of lugol and cover with a coverslip, leaving it to rest for 20 minutes or until the starch shows a colour that can vary from blue to violet-blue.

To identify lipids, as well as starch, a batch of 20 seeds was randomly selected using the sundan III method proposed by Ventrella et al. (2013). To carry out the test, the colouring solution had to be prepared, in which 3g of sudan III or IV reagent was dissolved in 100 mL of 70% ethanol, heated to boiling point and kept at rest until the solution cooled. Longitudinal and transverse sections of T. gardneriana Wedd. fruit were then placed in a petri dish lined with 9 cm diameter filter paper saturated with distilled water and placed in the freshly filtered sudan solution for 15 to 30 minutes in a closed container to prevent the solvent from evaporating and precipitates from forming. The sections were then washed quickly in 70% ethanol and then in water to mount the slides. The slides mounted in hydrated medium were placed in petri dishes lined with filter paper saturated with water. Drops of freshly filtered sudan were added to the sections and the dish was covered for 30 to 60 minutes to prevent the solvent from evaporating. They were then gently washed in a petri dish filled with water and dried at room temperature.

Both data were analysed daily for 31 days and the results were recorded using microscopic photographs using Leica DM 500 equipment with an attached microcomputer.

The morphological classification of the seeds was carried out with the aid of a 0.01m precision caliper, detecting the coefficient of variation, maximum and minimum value, mean and standard deviation by measuring the length, width and thickness of 100 embryos. Afterwards, 200 seeds from the batch of embryos stored in the laboratory were randomly selected for the germination test, in an attempt to check the germination percentage of the tree in relation to time. The seeds were distributed in ten 300ml transparent plastic containers, with a lid lined with 9cm diameter filter paper, moistened with 10ml of water, with twenty embryos in each replicate, analysed every two days and kept in a B.O.B. germination chamber, considering as germinated those individuals that develop a radicle.

The same methodology was used to detect photoblastism and viability as for the germination test. To determine the photoblastic group to which T. gardneriana Wedd. belongs, five flasks remained in the presence of light and the other five were covered with aluminium foil to eliminate any light source. For the viability analysis, which tested the purple colour of the living tissues present in the seeds, the five replicates were soaked in water and replaced after 24 hours with a 1% tetrazolium solution, which remained for a further 24 hours. The embryos were then cut lengthways, covering the apex and base of the embryo.

To produce Triplaris gardneriana Wedd. seedlings, seeds were extracted from fruit taken from adult trees located in the urban area of the municipality of Goiânia, Goiás. The seeds were germinated in laboratory conditions and then planted in plastic bags (15x12 cm) containing organic substrate. The plants received periodic irrigation and were kept under 50% shade in an area of the State University of Goiás (UEG), Morrinhos Unit.

After 180 days, 10 plants were randomly sampled to assess their morphology, height and diameter. Morphological characterisation was carried out using analytical morphology keys and height and diameter were measured using a caliper.

CHAPTER 3

RESULTS AND DISCUSSION

The T. gardneriana Wedd. species has a triangular achene fruit with a brown, smooth, glabrous and shiny epicarp (Figure 2-A). The fruit is dry and indehiscent, with a non-succulent endocarp and pericarp. The epidermis of the fruit of this species is thin and the seed is attached to the thin pericarp of the fruit.

It was observed that the seeds of the antwood (T. gardneriana) are bitegumentary, i.e. their envelope is divided into a testa and a tegmen (Figure 2-B,L). The forehead consists of the thicker outer part and the tegmen can be seen below the forehead, in the innermost light brown part surrounding the embryo. Thus, the hilum is not developed, as it often occurs in ategumentary ovules, which develop a scar located in the region separating the cotyledons of eudicots. It was found that the fruit of T. gardneriana has neither a raphe nor a micropyle, as the dispersing structure is a diaspore.

The embryo is cotyledonary, the cotyledons being white in colour at the start of germination and reddish yellow near the appearance of the coleoptile on the surface, being dispersed in the central part, having a membranous consistency (Figure 2-D,K,L). The embryonic axis is located below the cotyledons, consisting of the hypocotyl and the apical meristem of the radicle (Figure 2-C).

The presence of ruminated endosperm was verified as a reserve tissue that disappears completely before the cotyledons detach from the tegument tissues, corresponding to the evolution and development of the seedling, which occurs rapidly between 4 and 6 days (Figure 2-E,G,H,I,J,K,L). The disappearance of the endosperm was observed from the fourth day onwards (Figure 2-E,F), with the appearance of small gelatinous cracks in the transverse and horizontal sections. On the following days (from the fifth to the ninth) (Figures 2-G,H,I,J,K), these cracks became larger and larger until the structure completely disappeared, allowing room for the cotyledon to expand. Appezzato-da-Glória and collaborators (2013) state that this factor comes from the consumption of the endosperm by the embryo throughout its development, since in seeds with large cotyledons the endosperm plays the role of accumulating reserves for the development of the embryo.

Germination, as described by Andrade et al. (2008), is a process characterised by the breaking of the integumentary tissue by the primary root. In T. gardneriana Wedd. this phenomenon began seven days after the start of the germination process, in which the seedlings proved to be epigeal with cotyledons above ground level and phanerocotyledonous with foliaceous cotyledons free of seminal remains (Figure 3-B). The primary roots are glabrous, white in colour and cylindrical in shape (Figure 3-A). As the hypocotyl develops and shows a reddish-yellow colour, two chlorophyllous membranous cotyledons appear, partially enveloped by the tegument, around the 14th day (Figure 3-B).

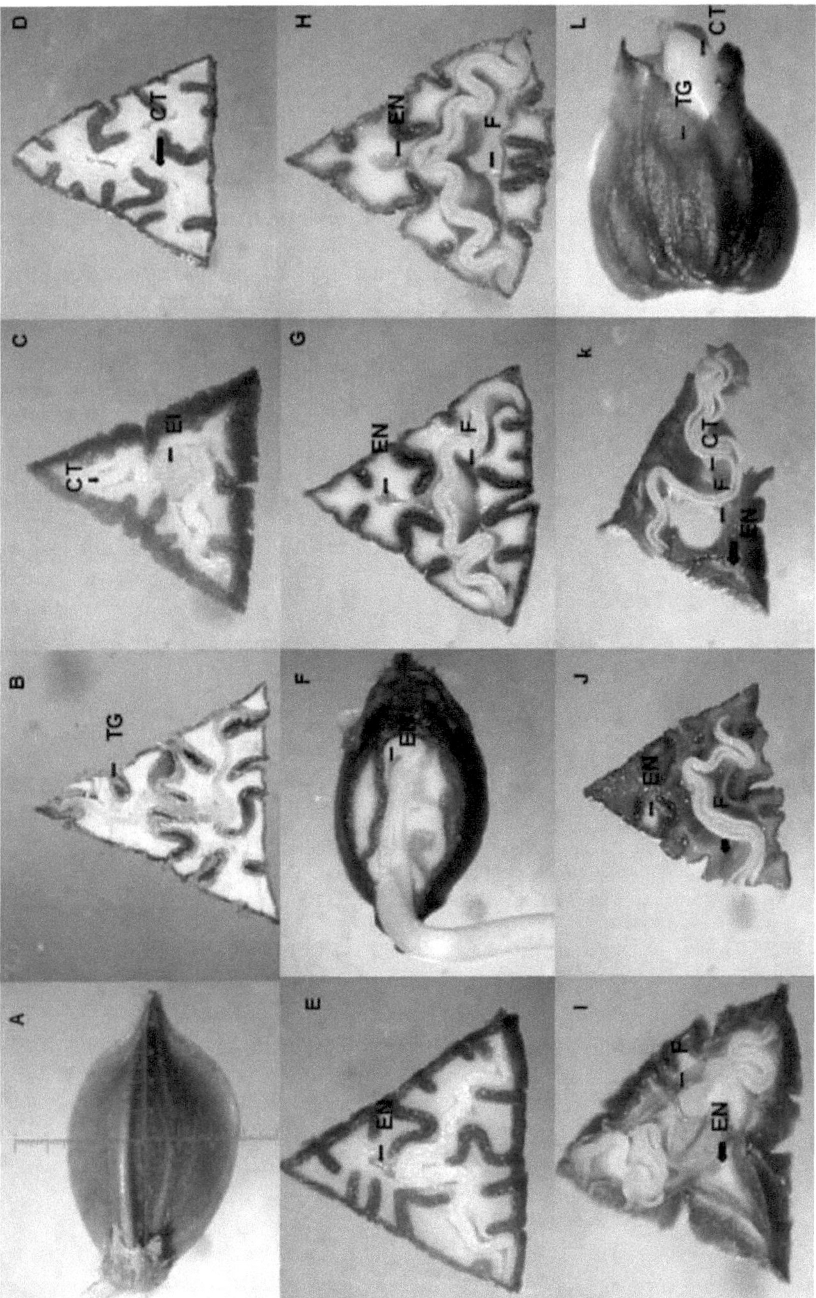

Figure 2 - General fruit morphology of Triplaris gardneriana Wedd.: a (external morphology of the fruit), b (tegmen = TG), c (embryonic axis = EI, cotyledons = CT)

21

, d (cotyledons = CT) , e (endosperm = EN), f (endosperm = EN), g (endosperm = EN, gelatinous fissure = F), h (endosperm = EN, fissure = F), i (endosperm = EN, fissure = F), j (endosperm = EN, fissure = F), k (endosperm = EN, fissure = F, cotyledons = CT), l (tegmen = TG, cotyledons = CT).

The cotyledons begin to open, with the tegument remaining attached to one of the cotyledonary leaves for around another 6 days and then falling off, along with the full opening of the cotyledonary leaves (Figure 3-C,D). Approximately 33 days after germination, the first leaves of the young plant (eophylls) appear (Figure 3-E).

At 53 days it was observed that the first leaves were simple, membranous, peninert and that the seedling still had cotyledons (Figure 3-F). The cotyledons remain for another month and then fall off, resulting in an 83-day cycle.

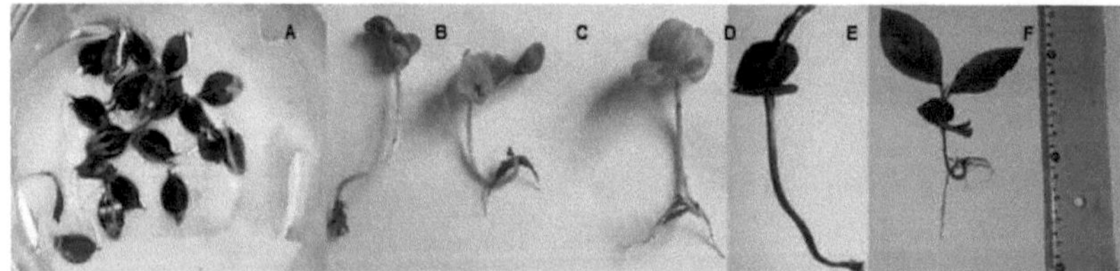

Figure 3 - General appearance of Triplaris gardneriana Wedd. seedlings: a (germination of pau-formiga seeds, highlighting the primary roots soon after the germination process), b (epigean and phanerocotyledonous seedling), c and d (development of the seedling and expansion of the cotyledons), e (emergence of the first leaf after 33 days), f (morphological characteristics of the seedling at 53 days after germination).

The average height of the seedlings 180 days after planting was 16.9 cm, with a standard deviation of 6.22 cm and a total range of 15 cm. The average stem diameter was 7.5 mm with a standard deviation of 1.8 mm. The plants had an average of eight leaves, the first of which (protophylls) had the same morphology as the metaphylls, with no heterophily (Figure 4).

Figure 4 - Morphological appearance of the seedlings at 180 days after planting Triplaris gardneriana Wedd,

In terms of morphological aspects, the leaves are simple with an elliptical limb, acuminate base and cuspidate apex. The youngest leaves have deciduous stipules that protect them until they expand (Figure 5). The young plant has alternate phyllotaxis, with a membranaceous consistency, a full edge and leaf and stem hairiness.

Figure 5 - Single leaf of a young Triplaris gardneriana Wedd. seedling, showing the morphological characteristics at 180 days after planting.

The venation is of the brochidodroma type, which is a special subtype of camptodroma venation, where the lateral veins are joined together by normally curved arches (Figure 6).

With regard to the type of stomata present in the species studied, it was found that the leaves are hypostomatic, with paracytic stomata. According to Appezzato-da-Gloria and Collaborators (2013), paracytic stomata have two subsidiary cells with their longitudinal axis parallel to the stomatal slit, as was found in the epidermis of the antwood leaf (Figure 6).

Stomata are responsible for gas exchange in plants and are therefore of great importance for the survival of terrestrial plants (MARTINS, 2009), which emphasises the importance of describing the stomata type of species endemic to the Cerrado.

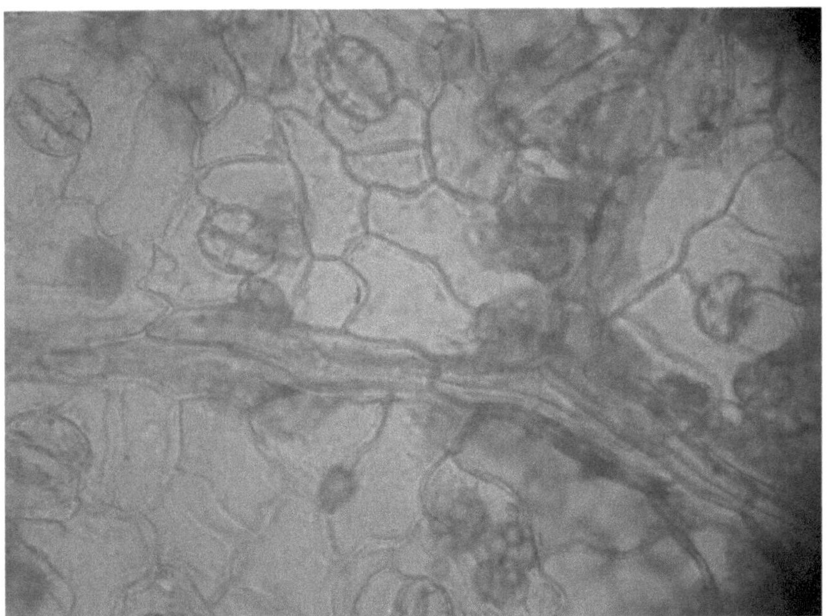

Figure 6- Paradermal section of the abaxial side of the leaf epidermis of Triplaris gardneriana Wedd. 400x magnification.

The description of the morphological characteristics of species provides biological data of great importance for studies aimed at taxonomy, favouring identification in the natural environment.

The shape of the leaf limb is one of the main characteristics analysed to describe species, due to the morphological diversity or variation that occurs between genera, families and species (SOUZA et. al. 2013). This extraordinary variation can also occur in plants of the same species, as was seen in the leaves of Triplaris gardneriana Wedd. sampled after three years.

The apex of the antwood leaf blade showed two morphological types, with a frequency of approximately 30% mucronate and 70% cuspidate, demonstrating that there is a natural variation in the morphology of Triplaris gardneriana leaves, in which the predominant apex type was cuspidate. The base showed a percentage variation of 70% for an obtuse base and 30% with a wedge-shaped base, thus characterising the species with obtuse base leaves. The edge was of the whole type, i.e. smooth and without any deformation on the blade. However, it has been observed that wavy limbs can appear on the same plant.

Leaf area is one of the main parameters of plant growth, due to its relationship with the plant's physiological processes (SILVA et. al., 2008). Moraes et. al., (2013) states that estimating or measuring leaf area is considered an important parameter for understanding the plant's interaction with the environmental factors involved, as well as the ecological evolution presented by the species.

Thus, by analysing the results obtained by estimating the leaf area of the Triplaris gardneriana Wedd. species using the length X width values and the actual leaf area, it was possible to obtain the following equation (Leaf area = -0.283 + 0.673*Length X Width), with a standard error of 11.549, an R of 0.970, and an R^2 of 0.941.

Using the equation obtained for this species, it is possible to estimate the leaf area without the use of measuring equipment, simply by replacing the value of x with the value multiplied by the greatest value of the leaf length by the greatest leaf width, obtaining the estimated leaf area. The relative values between the estimated leaf area and the observed values are presented, showing that the model presented explains the morphological and biometric variability of the antwood leaves. It shows a good fit between the observed and estimated values, but the error of the proposed estimate is not constant, suggesting the elimination of "outliers" from four observed values of x and y, namely: (83.79 - 56.25), (74.75 - 48.75), (66 - 27.5) and (67.85 - 42.5). This variation in the regression can be explained by leaf herbivory, bearing in mind that the leaves used for the study were collected at random.

After eliminating the outliers, the variances were stabilised, obtaining an R of 0.995, R2 of 0.991 and a standard error of 4.551, giving the following regression: (Leaf area = - 2.377 + 0.68739*Length x Width) (Figure 7). Analysing the difference between the actual leaf area values and the values estimated using the equation showed that there was no significant difference between them using the chi-squared test at the 5% level of statistical significance.

Due to the effectiveness, speed and precision of the results obtained, as well as the low cost of the experiment, the use of mathematical models, such as the linear regression adopted in this work, is suggested as a technique for obtaining leaf estimates of plant species such as Triplaris gardneriana Wedd.

The same is suggested by Lopes et. al. (2004), in their study with grapevines (Casta jaen), which shows that this model can be applied at any stage of the plant's biological cycle, with the same efficiency and precision as the use of devices.

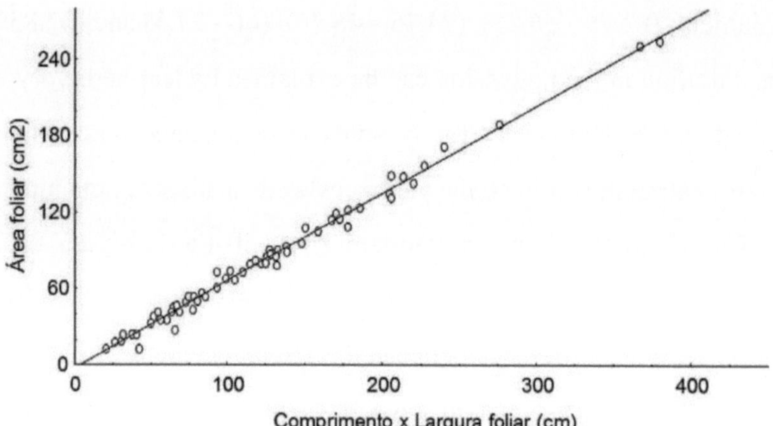

Figure 7 - Representation of the linear equation for obtaining the leaf area of the species Triplaris gardneriana Wedd. conducted at the State University of Goiás, Morrinhos Campus, Goiás, obtaining the following regression without outliers (Leaf area = -2.377 + 0.687 39* Length x Width).

The histochemical analysis of the T. gardneriana seed tissues showed the absence of lipid substances in the endosperm (Figure 8) and also in the embryo axis. According to Dietrich and Ribeiro (2015), some seeds have lipids stored in them to help develop the embryo during germination or the bud during sprouting, and the species studied did not show this behaviour. This means that the antwood depends on another source of energy and that lipids do not influence its physiological activity during germination.

In plants, the main function of lipids and starch found in seeds is to provide energy. These reserve substances can vary from one species to another, their proportion depending on the metabolic and physiological needs of the embryos (APPEZZATO-DA- GLORIA E COLABORADORES, 2013).

The characterisation of these compounds in native plants can contribute to studies aimed at storage in germplasm banks and seed viability.

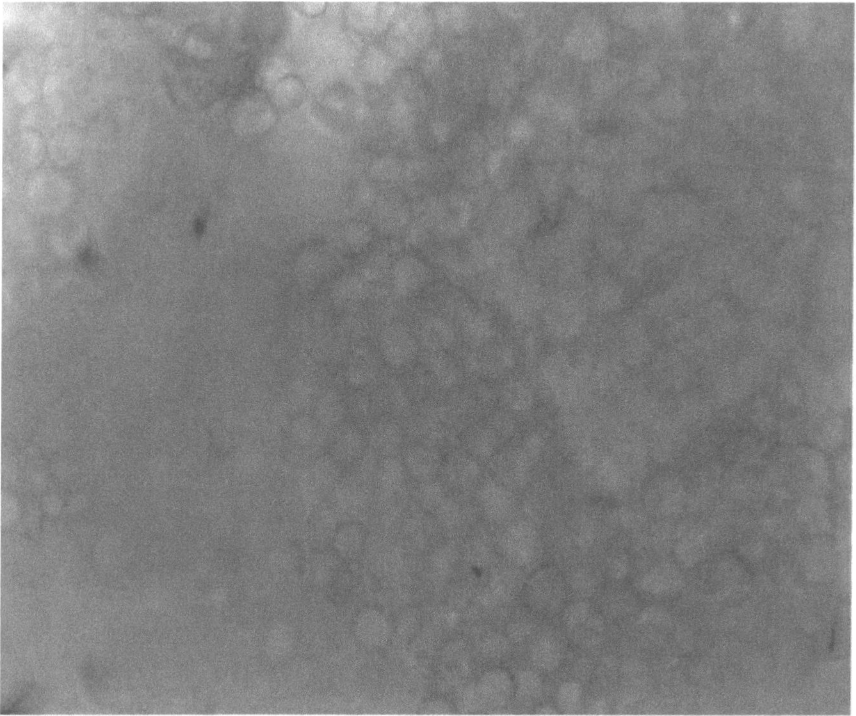

Figure 8 - Section of the endosperm of the seed of the species Triplaris gardneriana Wedd. conducted at the Universidade Estadual de Goiás, Câmpus - Morrinhos, Goiás, showing the absence of lipids in the cells. Magnification 1000 times.

In an attempt to detect the presence of starch in the tissues of pau-formiga (T. gardneriana) seeds, its presence was verified in the endosperm and embryonic axis (Figure 9). This shows that T. gardineriana uses starch as a reserve carbohydrate in its initial development phase, to carry out its physiological activities during germination.

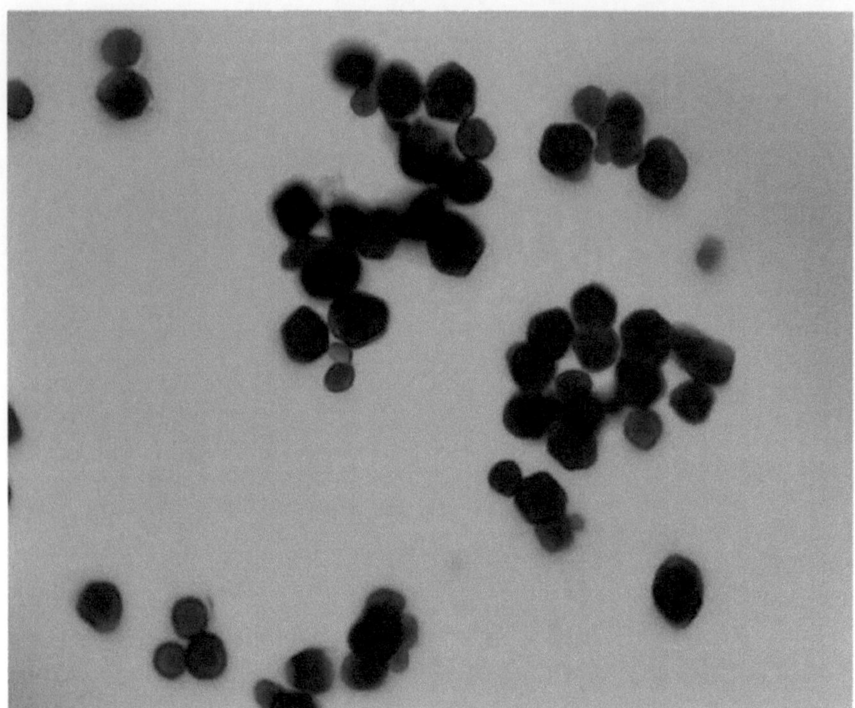

Figure 9 - Cut of the endosperm of the seed of the species Triplaris gardneriana Wedd. conducted at the State University of Goiás, Câmpus-Morrinhos, Goiás, observing the presence of starch in the cells using the lugol test. Magnification 1000 times.

Observing the results obtained from the biometric analysis, as Santos et al. (2007) argues that biometric characterisation allows species of the same genus to be differentiated in the field, and that class separations have been used to find the ideal class for the multiplication of different plant species, it can be seen that the seeds of Triplaris gardneriana Wedd. show variation in length from 10.2mm to 10.9mm, and in width and thickness from 0.3mm to 0.9mm (Table 1).

Table 1 - Dimensions and averages of antwood seeds (Triplaris gardneriana Wedd.).

Variables	Length (mm)	Width (mm)	Thickness (mm)

Maximum value	**10.9**	0.9	0,9
Average	10,633	0,537	0,633
Minimum value	10,2	0.3	0.3
CV{%)	2.20	36,49	30,17
Standard deviation	±0.234	± 0.196	± 0,191

CV= Coefficient of variation

And length, comprising 52% of the sample, was distributed in three biometric classes, with the greatest grouping among individuals with dimensions of 10.7 mm to 20.0 mm, followed by the 10.4 mm - 10.7 mm class with 31% (Figure 10A). While width remained highly concentrated among embryos in the 0.4 mm - 0.7 mm class with 47% (Figure 10B), and thickness with 50% and 44% in the 0.7 mm - 10.0 mm and 0.4 mm - 0.7 mm classes (Figure 10C).

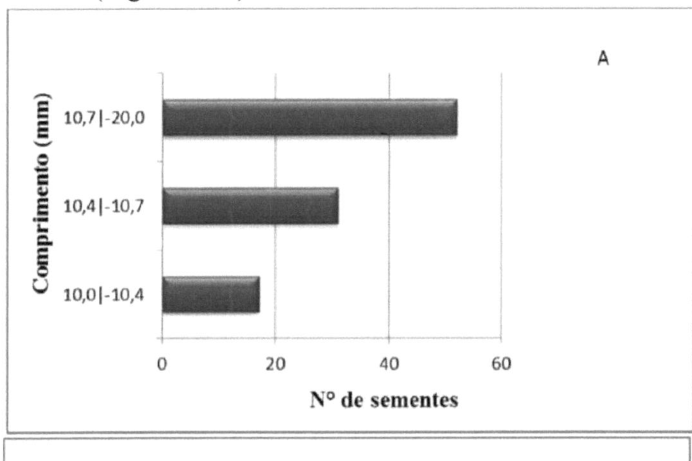

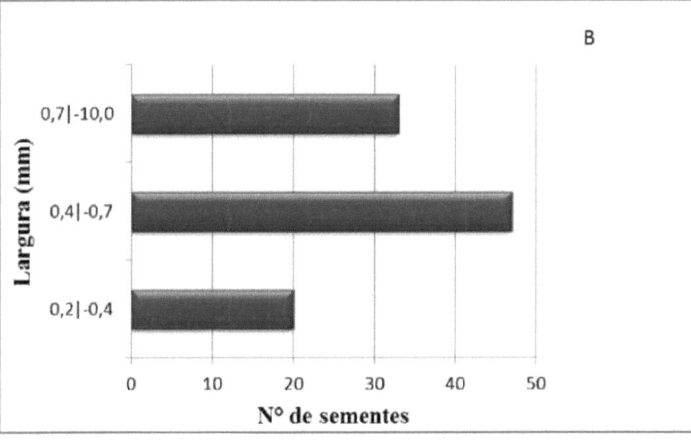

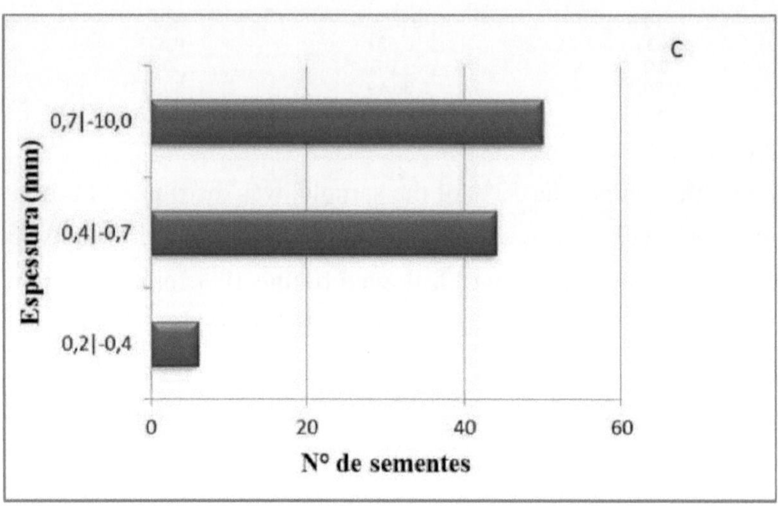

Figure 10 - Distribution of length classes (A), width (B) and thickness (C) of the seeds of the tree Triplaris gardneriana Wedd.

Antwood germination began three days after the start of the germination test, totalling nine days. The species thus showed a high germination percentage with rapid speed between the emergence of the radicles, as at three days the seeds showed 50% germination, at seven 71% and at nine 81.5% finalising the process (Figure 11). This proves the vigour and quality of the Triplaris gardneriana Wedd. embryos, as well as the probable lack of seed dormancy, giving the tree an ecological and biological advantage in its survival and performance in the biome, as well as the contribution it makes to the environment. Bearing in mind that reforestation and preservation programmes aimed at conserving native forest look for species that have quality covering the genetic, physiological and physical attributes of the seeds, it is performance that denotes the dynamic aspects of the implementation and completion of healthy seedlings, as high quality seeds perform better than those of lower quality (DELOUCHE, 2005).

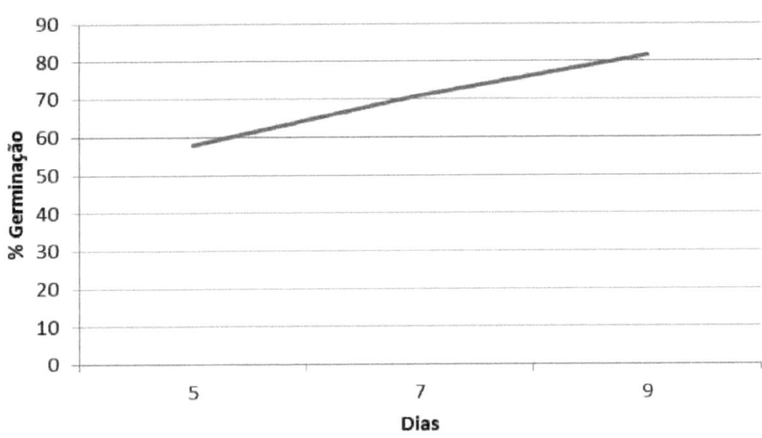

Figure 11 - Germination percentage (%) of two hundred embryos analysed up to the end of the germination process at nine days of Triplaris gardneriana Wedd. commonly known as pau-formiga.

The antwood has met the requirements demanded by environmentalists, because in addition to all the characteristics mentioned above, it is also a neutral photoblastic species with good viability.

Semantically, neutral photoblastism can be verified, since Triplaris gardneriana Wedd. shows good germination behaviour in both the light and dark states, with an insignificant difference in the amount of germination. In which 95% germinated in the light and 90% in its absence (Figure 12), which according to Silva, et al., (2001) affirms its ability to germinate in full sun or within closed forest formations.

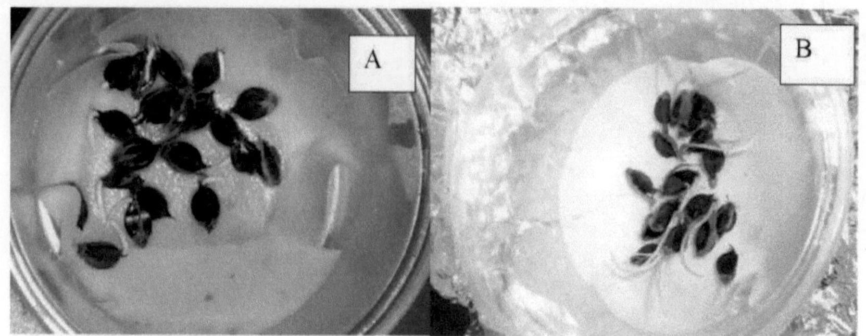

Figure 12 - Morphological characteristics of the seedlings from the photoblast test of T. gardneriana Wedd. seeds. (A) Germinated seedling in the light; (B) Germinated seedling in the dark 10 days after the start of the test.

With regard to viability, the antwood shows a high content of viable seeds, as can be seen in the samples (Figure 13), which show 7%, 8%, 8%, 9% and 10% of live embryos. In other words, the seeds of the species behaved as neutral photoblasts with 95% germination in the presence of light and 90%, showing no statistical significance at the 5% probability level using the t-test (Figure 13).

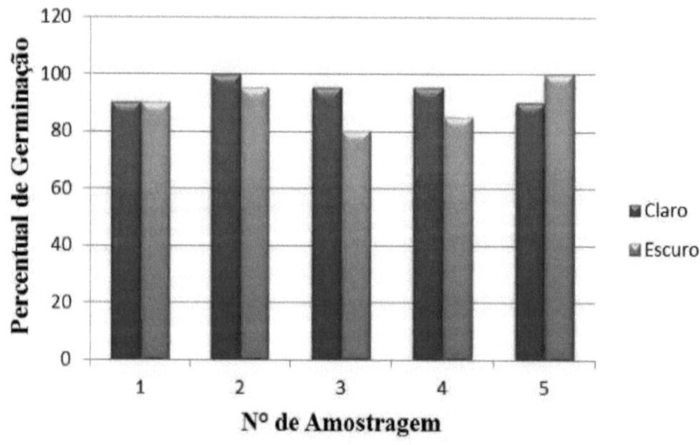

Figure 13 - Percentage of germination in the photoblast test of T. gardneriana Wedd. seeds based on the analysis of 100 embryos distributed and evaluated in five replicates in the light and five in the dark.

The tetrazolium test showed that an average of 92 per cent of the seeds were viable, a result consistent with the germination values.

CHAPTER 4

FINAL CONSIDERATIONS

The morphological characterisation of the Triplaris gardneriana Wedd. diaspores showed that the achenes are bitegumentary with rumen endosperm and their seed reserve material is starchy. It has been shown that the embryo is of the mature-eutrophic type, where the root and leaf structure are already pre-formed in the diaspore, in which the elongation of the embryo is sustained by the mobilisation of reserves located in the endosperm. Germination of the species is epigean and phanerocotyledonous with photosynthesising membranous cotyledons. The leaves of young plants are hypostomatic, typical of mesophytic plants, with paracytic stomata, a mucronate apex, obtuse base and entire edge.

The equation for estimating leaf area using linear measurements showed a good fit in relation to real data and could replace the use of equipment for this purpose. This is because the species, in its early stages of development, has morphological leaf characteristics similar to adult plants. These characteristics can help identify plants of this species while they are still in the vegetative stage.

The species has a high germination rate and a fast germination rate, as after nine days it had a germination rate of 81.5%, with neutral photoblastic seeds with 95% germination in the presence of light and 90% in its absence, so the species has a high percentage of viable embryos corresponding to 85%, 90%, 90%, 95% and 100%. These factors demonstrate its physiological and germination quality through uniformity, speed and quality in the germination process and later in the production of seedlings, making it an ideal plant to be used in reforestation, restoration, conservation, propagation and tree development projects in the cerrado.

CHAPTER 5

BIBLIOGRAPHICAL REFERENCES

ANDRADE, D.A.V.; ORTOLANI, F.A.; MORO, J.R.; MORO, F.V. Morphological aspects of fruits and seeds and cytogenetic characterisation of Crotalaria lanceolata E. Mey. (Papilionoideae - Fabaceae). Acta Bot. Bras. v. 22, n.3, p. 621-625, 2008.

APPEZZATO-DA-GLORIA, B.; CARMELLO-GUERREIRO, S.M. Anatomia vegetal. 3.ed., rev. and expanded: Viçosa, MG. Ed.UFV, 2013. 404p.

ARAÚJO-NETO, J.C.; AGUIAR, I.B.; FERREIRA, V.M.; PAULA, E.C. Morphological characterisation of fruits and seeds and post-seedling development of monjoleiro (Acacia polyphylla DC.). Revista Brasileira de Sementes, v. 24, n.1, p.203-211, 2002.

BASTOS, L.A.; FERREIRA, I.M. Phytophysiognomic compositions of the Cerrado biome: a study of the Vereda subsystem. Espaço em Revista, v. 12, n. 2, p. 97 - 108, 2010.

BATISTA, J.G.F.P. The importance of the world's biomes: and the cerrado in the Brazilian context. In: 10th NATIONAL MEETING ON TEACHING PRACTICE IN GEOGRAPHY, 2009, Porto Alegre. Proceedings of the 10th National Meeting of Teaching Practice in Geography, 2009. 11p.

BENINCASA, M.M.P. Analysing plant growth (basics).Jaboticabal: Funep, 2003.41p.

CASTRO, A.S.; CAVALCANTE, A. Flores da caatinga. Campina Grande: Instituto Nacional do Semiárido, 2010.116p.

CRUZ, C.C.; SOLANO, E. Familia Polygonaceae. Flora del bajío y de regiones adyacentes, 2008.51p. Mexico: Volume 153.

DÁTTILO, W.; ELISABETE DA COSTA MARQUES, E.C.; FALCÃO, J.C.F.; MOREIRA, D.D.O. Mutualistic interactions between ants and plants. EntomoBrasilis, v.2, n.2, p.32-36, 2009.

DELOUCHE, J.C. Seed Quality and Performance.2005.

DIETRICH, S.M.; RIBEIRO, R.C.L.F. Reserve carbohydrates in higher plants and their importance for humans. Accefyn, v.16, n.61, p.65-71, 2015.

DINIZ, B.P.C. O Grande cerrado do Brasil central: geopolitics and economy. 2006. 231f. Dissertation (Postgraduate) - USP/FFLCHDG, São Paulo,

2006.

DURIGAN, G.; MELO, A.C.G.; MAX, J.C.M.; VILAS BOAS, O.; CONTIERI, W.A.; RAMOS, V.S. Manual for the recovery of cerrado vegetation. 3.ed. rev. and current. São Paulo : SMA, 2011.19 p.

EMBRAPA. Cerrado Biome Biodiversity Conservation and Management Project (CMBBC): Final Product Report, p.7-11, 2005.

FELFILI, J.M.; RIBEIRO, J.F.; FAGG, C.W.; MACHADO, J.W.B. Cerrado: manual for recovering gallery forests. Planaltina: Embrapa Cerrados, 2000. 45p.

HADDAD-JUNIOR, V.H.; BICUDO, L.R.H.; FRANSOZO, A. The Triplaria tree (Triplaris spp) and Pseudomyrmex ants: a symbiotic relationship with risks of attack for humans. Revista da sociedade brasileira de medicina tropical, V.42, N.6, P.727-729, 2009.

JUDD, W. S.; CAMPBELL, C. S.; KELLOGG, E. A.; STEVENS, P. F.; DONOGHUE, M. J. Plant Systematics: a Phylogenetic Approach. 3rd ed. - Porto Alegre: Artmed, 2009. 632p.

LORENZI, Harri. Árvores brasileiras: Manual de identificação e cultivo de plantas arbóreas nativas do Brasil. 2ª Ed. Nova Odessa: Editora Plantarum, 2002. vol.2, 384 p.

LORENZI, H. Árvores brasileiras: manual de identificação e cultivo de plantas arbóreas nativas do Brasil. 1. ed., Nova Odessa, SP: Ed. Plantarum, 2009. 384p.

LORENZI, H. Árvores brasileiras: manual de identificação e cultivo de plantas arbóreas nativas do Brasil. 4. ed. v.2, Nova Odessa, SP: Ed. Plantarum, 2013. 384p.

LORENZI, H. Árvores brasileiras: manual de identificação e cultivo de plantas arbóreas nativas do Brasil. 6. ed., Nova Odessa, SP: Ed. Plantarum de Estudos da Flora, 2014. 384p.

LOPES, C.M.; ANDRADE, I.; PEDROSO, V.; MARTINS, S. Empirical models for estimating vine leaf area in the Jaen variety. Ciência Téc. Vitiv. V.19, n.2, p.61-75, 2004.

MACÊDO, N.A. Manual of techniques in plant histology. State University of Feira de Santana - UEFS, 1997. 90p.

MARTINS, G.A. Evaluation of stomata characteristics in jatoba (Hymenaea courbaril L.) using geostatistics. Lavras: UFLA,

2009.73p.

MENDONÇA, A.V. R.;COELHO, E.A.;SOUZA, N.A.;BALBINOT, E.;SILVA,R.F.;BARROSO, D.G. Efeito da hidratação e do condicionamento osmótico em sementes de pau-formiga. Revista Brasileira de Sementes, v. 27, n. 2, p.111-116, 2005.

MINISTRY OF THE ENVIRONMENT. Biodiversity of the Cerrado and Pantanal: priority areas and actions for conservation. Brasília: MMA, 2007. 540 p.

MINISTRY OF THE ENVIRONMENT. Action Plan for the Prevention and Control of Deforestation and Burning in the Cerrado - PPCerrado. Brasília: MMA, 2009.152p.

MINISTRY OF THE ENVIRONMENT. National Programme for the Conservation and Sustainable Use of the Cerrado Biome: Sustainable Cerrado Programme. Brasília: MMA, 2003.56p.

MINISTRY OF THE ENVIRONMENT. National Programme for the Conservation and Sustainable Use of the Cerrado Biome. Brasília: MMA, 2006. 50p.

MORAES, L.; SANTOS, R.K.; WISSER, T.Z.; KRUPEK, R.A. Evaluation of leaf area from simple linear measurements of five plant species under different light conditions. R. Bras. Biociências, Porto Alegre, v. 11, n. 4, p. 381-387, 2013.

MYERS, N.; MITTERMEIER, R.A.; MITTERMEIER, C.G.; FONSECA, G.A.B; KENT, J. Biodiversity hotspots for conservation priorities. Nature, v.403, n.24, p.853-858, 2000.

OLIVEIRA, P.S.; MARQUIS, R.J. The cerrados of Brazil: ecology and natural history of a neotropical savanna. Columbia University Press, 2002. 409p.

PESSOA, O.A.A. Temporal evolution of the spectral behaviour of a burnt area in a Cerrado grassland formation. 2014.105f. Dissertation (Master's) - UNB, Brasília, 2014.

PINHEIRO, M.R.; KURY, K.A. Environmental conservation and basic concepts of ecology. Boletim do Observatório Ambiental Alberto Ribeiro Lamego, Campos dos Goytacazes: Rio de Janeiro v.2, n.2, 2008.14p.

REIS, J. S. ; SILVA, D. A. ; CHAVES FILHO, J. T. Morphological description of seedlings of the tree species Triplaris gardneriana (Pau-Formiga). In: CONGRESSO FLORESTAL NO CERRADO & 3 SIMPÓSIO SOBRE EUCALIPTOCULTURA, 2013, Goiânia. Proceedings of the Forestry Congress in the Cerrado & 3 Symposium on Eucalyptus Culture, 2013. p. 156-157, 2013.

REIS, J. S.; CHAVES FILHO, J. T. Evaluation of germination and photoblastism in seeds of Triplaris gardneriana Wedd. (antwood). In: XI LATIN AMERICAN CONGRESS OF PHOTOGRAPHY , LXV CONGRESSO NACIONAL DE BOTÂNICA E XXXIV ENCONTRO REGIONAL DE BOTÂNICOS MG, BA, ES, 2014, Salvador. Proceedings of the XI LATIN AMERICAN CONGRESS OF BOTANICA, LXV NATIONAL CONGRESS OF BOTANICA AND XXXIV REGIONAL MEETING OF BOTANICALS. 65CNBot, 2014.

RIBEIRO, P.R.C.C.; RIBEIRO, J.J.; NETO, A.S.; ROCHA, J.R.P.;

CORTE, I. Methods for recovering riparian forest as a proposal for recovering springs in the cerrado. Enciclopédia biosfera, Centro Científico Conhecer, Goiânia, v.8, n.15, 2012. 17p.

SANTIAGO, J.; JUNIOR, M.C.S.; LIMA, L.C. Phytosociology of tree regeneration in the Pitoco Gallery Forest (IBGE-DF), six years after accidental fire. Scientia Forestalis, n. 67, p.64-77, 2005.

SANTOS, R.M.; VIEIRA, F.A. Floristic similarity between dry forest and gallery forest formations in the sapucaia municipal park, Montes Claros-MG. Electronic scientific journal of forestry engineering, ano. IV, n.07, 2006. 10p.

SANTOS-DINIZ, V.S.; SOUSA, T.D. Floristic and phytosociological survey of dry semi-deciduous forest in a legal reserve area in the municipality of Diorama, western Goiás, Brazil. Enciclopédia biosfera, Centro Científico Conhecer, Goiânia, v.7, n.12, p.1- 17, 2011.

SANTOS, F.S. Biometry, germination and physiological quality of Tabebuia chrysotricha (Mart. Ex A. Dc.) Standl. seeds from different matrices.2007.48f. Dissertation (Master's) - UEP-FCAV, Jaboticabal, 2007.

SILVA, D.B.; LEMOS, B.S. Plantas da área verde da super quadra norte 416- Brasília. DF. Brasília: Embrapa Recursos Geneticos e Biotecnologia, 2002. 147 p.

SILVA, M.C.C.;NAKAGAWA, J.;FIGLIOLIA, M.B. Influence of temperature, light and water content on the germination of Schinus terebinthifolius Raddi-Anacardiaceae (Red Aroelra) seeds. Rev. Inst., Flor., São Paulo, v. 13, n. 2, p. 135-146, 2001.

SILVA, L.A.G.C. Biomes present in the state of Tocantins. Legislative Consultancy: Technical Note, 2007. 9p.

SILVA, N.L.A.; MIRANDA, F.A.A.; CONCEIÇÃO, G.M. Phytochemical Screening of Cerrado Plants, from the Municipal Environmental Protection Area of the

Inhamum, Caxias, Maranhão. Scientia Plena, v.6, n.2, p.01-17, 2010.

SILVA, S.H.M.G.; LIMA, J.D.; BENDINI, H.N.; NOMURAL, E.S.;

MORAES, W.S. Estimation of anthurium leaf area using regression functions. Ciência Rural: Santa Maria, v.38, n.1, p.243-246, 2008. Santa Maria, v.38, n.1, p.243-246, 2008.

SILVA-JUNIOR, M.C.S. 100 Trees of the Cerrado sentido restrito: Field Guide. Brasília: Cerrado Seed Network, 2012.304p.

SILVEIRA, E.P. Floristics and structure of cerrado sensu stricto vegetation on indigenous land in the north-west of the state of Mato Grosso. 2010. 62f. Dissertation (Postgraduate) - UFMG/FEF, Cuiabá, 2010.

SOUZA, V.C.; FLORES, T.B.; LORENZI, H. Introduction to botany: morphology. São Paulo: Instituto Plantarum de Estudo a Flora, 2013. 222p.

VENTRELLA, M.C.; ALMEIDA, A.L.; NERY, L.A.; COELHO, V.P.M. Histochemical methods applied to seeds. Viçosa: Ed. UFV, 2013. 40p.

WARD, P.S. Systematics, biogeography and host plant associations of the Pseudomyrmex viduus group (Hymenoptera: Formicidae), Triplaris- and Tachigali-inhabiting ants. Zoological Journal of the Linnean Society, v.126, p.451-540, 1999.

I want morebooks!

Buy your books fast and straightforward online - at one of world's fastest growing online book stores! Environmentally sound due to Print-on-Demand technologies.

Buy your books online at
www.morebooks.shop

Kaufen Sie Ihre Bücher schnell und unkompliziert online – auf einer der am schnellsten wachsenden Buchhandelsplattformen weltweit! Dank Print-On-Demand umwelt- und ressourcenschonend produziert.

Bücher schneller online kaufen
www.morebooks.shop

MIX
Papier aus verantwortungsvollen Quellen
Paper from responsible sources
FSC® C105338

Printed by Books on Demand GmbH, Norderstedt / Germany